PARTS OF A PLANT

SEEDS

A Crabtree Roots Book

ALICIA RODRIGUEZ

Crabtree Publishing
crabtreebooks.com

School-to-Home Support for Caregivers and Teachers

This book helps children grow by letting them practice reading. Here are a few guiding questions to help the reader with building his or her comprehension skills. Possible answers appear here in red.

Before Reading:

- What do I think this book is about?
    - *This book is about seeds.*
    - *This book is about what seeds look like.*

- What do I want to learn about this topic?
    - *I want to learn what a seed is.*
    - *I want to learn what a seed looks like.*

During Reading:

- I wonder why...
    - *I wonder why some seeds need soil.*
    - *I wonder why some seeds are little.*

- What have I learned so far?
    - *I have learned that seeds can be big.*
    - *I have learned that seeds need water.*

After Reading:

- What details did I learn about this topic?
    - *I have learned that seeds can be different sizes.*
    - *I have learned that not all seeds need soil.*

- Read the book again and look for the vocabulary words.
    - *I see the word **sunlight** on page 8 and the word **soil** on page 10. The other vocabulary word is found on page 14.*

This is a **seed**.

Some seeds are big.

Some seeds are little.

Most seeds need **sunlight**.

Most seeds need **soil**.

All seeds need water.

Word List

Sight Words

a
are
is
most
need
some
this

Words to Know

seed

soil

sunlight

24 Words

This is a **seed**.

Some seeds are big.

Some seeds are little.

Most seeds need **sunlight**.

Most seeds need **soil**.

All seeds need water.

PARTS OF A PLANT
SEEDS

Written by: Alicia Rodriguez
Designed by: Rhea Wallace
Series Development: James Earley
Proofreader: Ellen Rodger
Educational Consultant: Marie Lemke M.Ed.
Production manager: Candice Campbell

Photographs:
Shutterstock: SingKham: cover; Joaquin CorbaianP: p. 1; menic183: p. 3, 14; Ir.S: p. 5; Dipak Shelare: p. 7; teerayuth oanwong: p. 8, 14; DibasUA: p. 11, 14; amenic181: p. 12

Crabtree Publishing

crabtreebooks.com 800-387-7650

In Canada: We acknowledge the financial support of the Government of Canada through the Canada Book Fund for our publishing activities.

Printed in the U.S.A./CP052026

Library and Archives Canada Cataloguing in Publication

Title: Seeds / Alicia Rodriguez.
Names: Rodriguez, Alicia (Children's author), author.
Description: Series statement: Parts of a plant | "A Crabtree roots book".
Identifiers: Canadiana (print) 20210176091 | Canadiana (ebook) 20210176105 | ISBN 9781427140685 (hardcover) | ISBN 9781427140746 (softcover) | ISBN 9781427133304 (HTML) | ISBN 9781427133908 (EPUB) | ISBN 9781427140807 (read-along ebook)
Subjects: LCSH: Seeds—Juvenile literature.
Classification: LCC QK661 .R63 2022 | DDC j575.6/8—dc23

Published in Canada
Crabtree Publishing
616 Welland Avenue
St. Catharines, Ontario
L2M 5V6

Published in the United S
Crabtree Publishing
347 Fifth Avenue
Suite 1402-145
New York, NY 10016

Hardcover	978-1-4271-4068-5
Paperback	978-1-4271-4074-6
Ebook (pdf)	978-1-4271-3330-4
Epub	978-1-4271-3390-8
Read-along	978-1-4271-4080-7
Audio book	978-1-4271-3914-6

Library of Congress Cataloging-in-Publication Data

Names: Rodríguez, Alicia, 1992- author.
Title: Seeds / Alicia Rodriguez.
Description: New York, NY : Crabtree Publishing Company, [2022] | Series: Parts of a plant - a Crabtree roots book | Includes index.
Identifiers: LCCN 2021014298 (print) | LCCN 2021014299 (ebook) | ISBN 9781427140685 (hardcover) | ISBN 9781427140746 (paperback) | ISBN 9781427133304 (ebook) | ISBN 9781427133908 (epub) | ISBN 9781427140807
Subjects: LCSH: Seeds--Juvenile literature. | Plants--Juvenil literature.
Classification: LCC QK661 .R634 2022 (print) | LCC QK661 (ebook) | DDC 581.4/67--dc23
LC record available at https://lccn.loc.gov/2021014298
LC ebook record available at https://lccn.loc.gov/2021014299